Science of Little Round Things

Also by Benita L. Epstein

Suture Self: Cartoons for Doctors and Patients
(McFarland, 1999)

Interlibrary Loan Sharks and Seedy Roms: Cartoons from Libraryland
(McFarland, 1998)

Science of Little Round Things

Cartoons About Scientists

by

BENITA L. EPSTEIN

McFarland & Company, Inc., Publishers
Jefferson, North Carolina, and London

Library of Congress Cataloguing-in-Publication Data

Epstein, Benita L., 1950–
Science of little round things : cartoons about scientists /
by Benita L. Epstein.
p. cm.
ISBN 0-7864-0807-3 (softcover : 55# alkaline paper) ♾
1. Scientists—Caricatures and cartoons. 2. American wit
and humor, Pictorial. I. Title.
NC1429.E68A4 2000
741.5'973—dc21 99-56292

British Library Cataloguing-in-Publication data are available

Manufactured in the United States of America

McFarland & Company, Inc., Publishers
Box 611, Jefferson, North Carolina 28640
www.mcfarlandpub.com

To Ted

Contents

Introduction

Ever since I became a cartoonist eight years ago I am often asked, "How long does it take to draw a cartoon?" I say it takes five minutes to sketch the scene, ten minutes to ink the final drawing and 20 minutes to add shading. That's a lie. To sift through my brain for the ideal funny joke and to perfect my style took over 40 years.

That's when my career really began, in the 1960s. I was a budding scientist then, gathering my first collection of earwigs, moths and katydids. Around this time I also drew my first cartoon character, The Destroyer, who was based on a popular wrestler. I still use this character in many of my background crowd scenes.

I kept an interest in both science and art, earning bachelor's and master's degrees in entomology. After graduating I worked in various research laboratories. I fed colonies of yellow fever mosquitoes blood from my forearm, isolated lung surfactant from human amniotic fluid, and performed experiments in molecular genetics on metabolic diseases. Little did I realize I would actually use the information gleaned from 20 years of titrating buffers, labeling eppendorf tubes, or growing cell cultures. I now use my original mosquito pupa picking pipet to dilute India ink for shading cartoons. The signature, BLE, my initials, that I wrote rapidly and indecipherably on a thousand bottles of solutions is my cartoon signature now. I knew someday I'd finally find a more pleasant use for my expert manual dexterity.

After 20 years conducting research I decided I had gathered enough funny material to become a professional cartoonist. I don't need to calibrate pH meters or adjust incubators anymore. I don't need to warm up spectrophotometers or balance tubes in an ultracentrifuge. Now I only need imagination and my favorite pen. I can take all those years and convert creative energy into a modest living. Thanks to the support of my husband, a biology professor, I am able to write funny gags while reclining on the couch, draw wacky cartoons and submit them to publishers.

The cartoons in this book were influenced by my background in entomology, ecology, and lab work. The other areas, physics, chemistry, cell biology, all the fields I studied but forgot, I classify as the science of "little round things." In fact, the first cartoon I ever sold was called "science of little round things."

At the end of this book is a section on another frequently asked question, "Where do you get your ideas?"

Benita Epstein
January 2000
Cardiff, California

The original versions of many of the cartoons in this book first appeared in *American Scientist, Natural History, Discover, Earthwatch, Air & Space* and *The Chronicle of Higher Education.* My work now appears in hundreds of other publications such as *The New Yorker, Barron's, USA Weekend, Wall Street Journal* and two other McFarland cartoon collections, *Suture Self: Cartoons for Doctors and Patients* and *Interlibrary Loan Sharks and Seedy Roms.*

"Help yourself. It's supposed to be very good for the rainforest."

"You have a Y chromosome. I like that in a man."

"Everything's DNA these days. Find me some."

"We're not lost. Here's a map of the human genome."

"No one understands me."

"No, I'm not the waiter. I'm the genetic engineer. How would you like your lamb?"

"This one is certainly not well-suited to its environment."

"Professor, wake up! It's spring. Your grant's over!"

"Check out Mr. Bigshot. Got his teeth whitened."

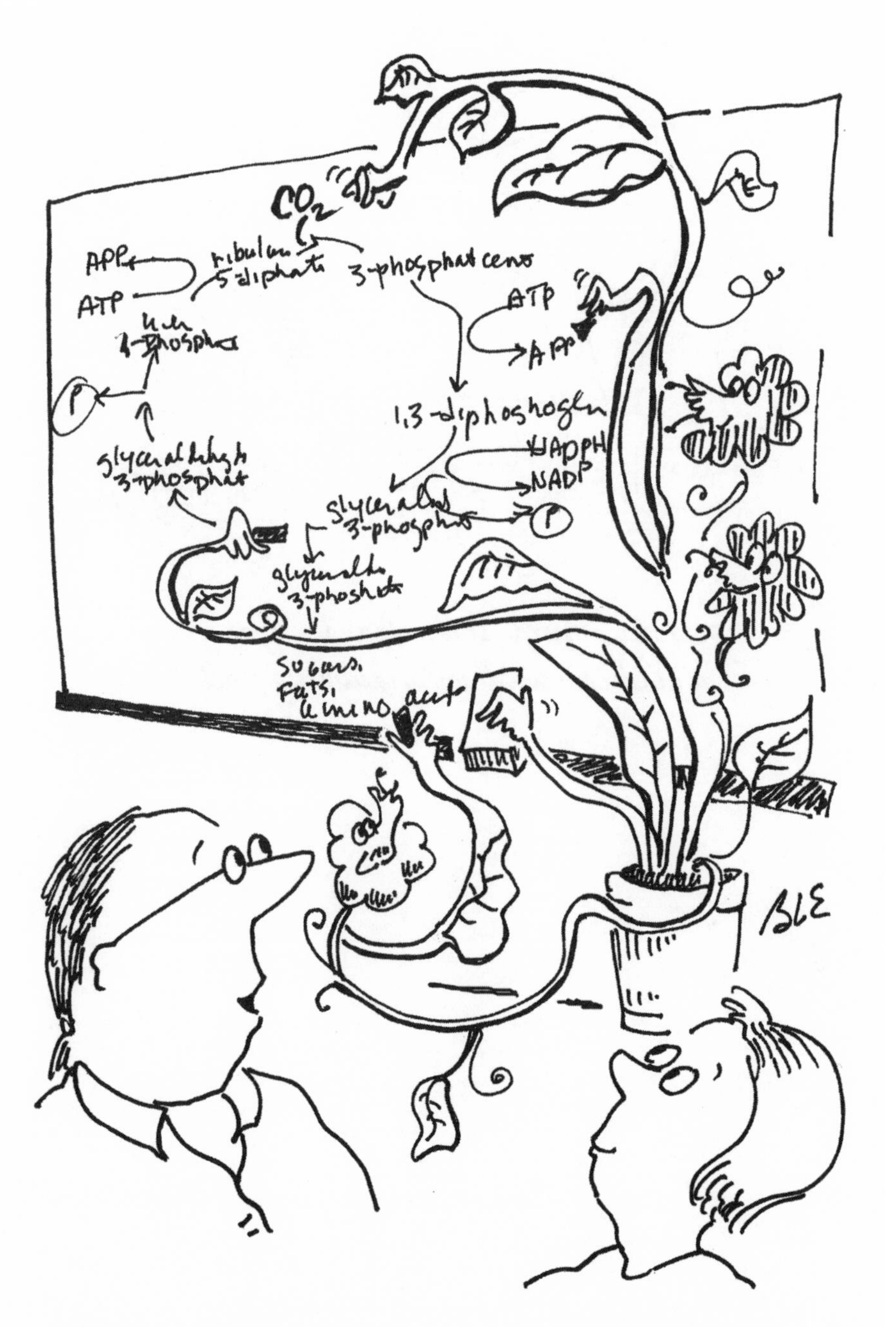

"That's so cool how they convert carbon dioxide to oxygen."

"Don't be absurd. If we cloned a hyena into a mad cow we'd get a laughing stock."

"The labs are back."

POSTER OF BEE GEES
GRANTS
WEIRD GUY FROM HIGH SCHOOL
TAX RECEIPTS
8th GRADE SCIENCE PROJECT
LOST IN SPACE

UNPUBLISHED AUTHORS OF THE BIG BANGS THEORY

THE RED SHIFT
THE PURPLE MUU-MUU

THEORIES
EXPOUNDED

"Do you have this grant in a larger size?"

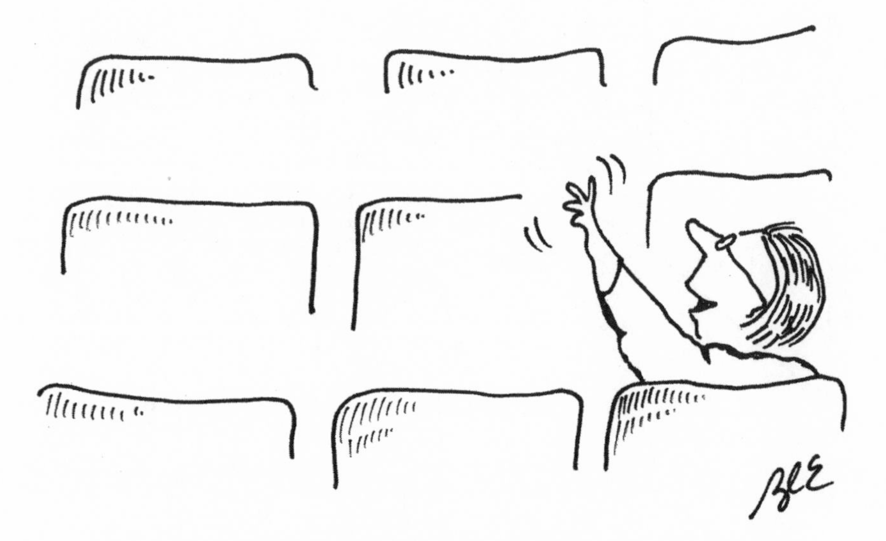

"The question raised was 'How did such a nice boy get such a big lab with all those awards?'"

DRUG
DE-TOX
ALCOHOL
DE-TOX
E-MAIL
DE-TOX

"What did you THINK the 'escape key' was for?"

"I'd like to help the environment. I'll buy this soap for $17.50."

"An ornithologist? Where are your offices?"

Complete Metamorphosis

"Day 573 of the project: Let's just say there are many, many, many, many ants and leave it at that."

AHH. THE SHELL OF MY FORMER SELF.
BLE

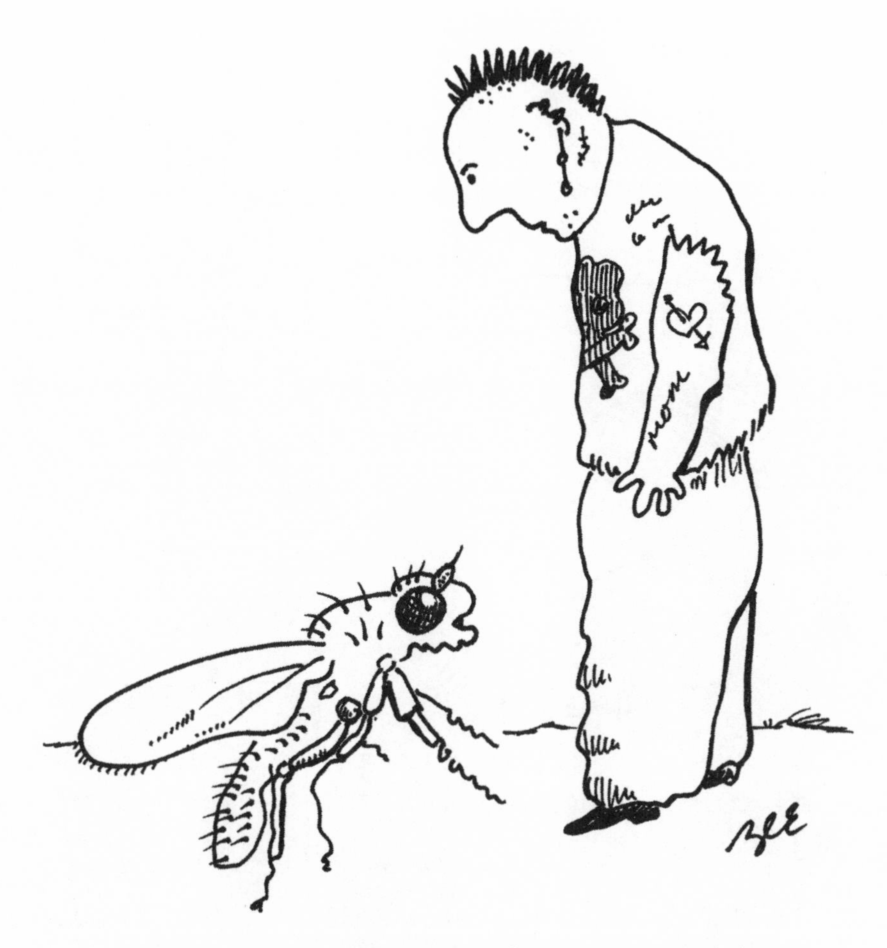

"Same as you. We metamorphose, mate and die."

"It's ten o'clock. Do you know where your research is headed?"

THE TAKE YOUR MOTHER TO WORK PROGRAM

MATHEMATICIAN'S INSOMNIA

"Thanks, but that's not wallpaper.
It's just a bunch of differential equations."

SAVE
THE
WHALES!
FREE
RADICALS
!!!

"Your genes need altering."

"You've already showed me how well you were trained."

"You can't keep running in here and demanding data every five years!"

"Oh, me? I gave a seminar, wrote a bunch of grants, did some experiments and now it's lunchtime."

"Finally! A man who's interested in my brain."

"He put the 'bop' in the suboperculum."

"Notice anything different?"

"This is the class Insecta. You want the class Arachnida down the hall."

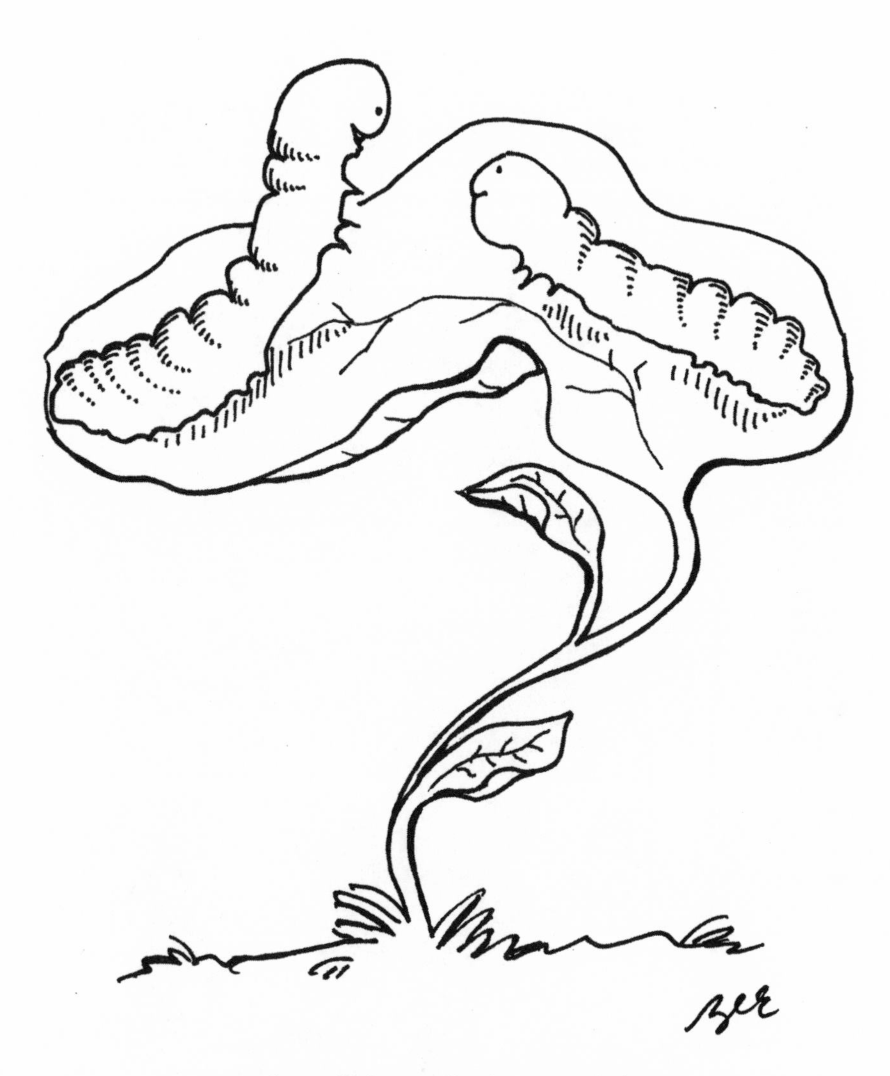

"What did you do for Earthday?"

"How much did you save?"

"They say that for every known species of insect there are ten more waiting to be discovered."

GENETIC
ENGINEERING
'N
THINGS

"Don't breathe a word to anyone, but I'm never leaving graduate school."

ONE OF MANY CALIFORNIA GENE POOLS

"And this one's for entering an animal facility under siege."

THE NEW GEOLOGICAL TIMETABLE
CENOZOIC
"JEOPARDY" INVENTED
FIRST SPORT UTILITY VEHICLE
BEN + JERRY'S ICE CREAM
MESOZOIC
PACKAGED LETTUCE
BARNEY
PALEOZOIC
EXPANSION OF PASTA
SNAIL MAIL BEGINS
PREHISTORIC Y2K BUG
PRECAMBRIAN
CELL
CELLULITE

"Can I take my break now?"

FIRST WORDS EVER SPOKEN BY A MOTHER

"All these years I've been meaning to ask . . . are you male or female?"

What phylogenetic relationship?
That was a million years ago."

"You call THAT a knack for capturing nature?"

NO MATTER WHERE, EVERYONE GRAVITATES TOWARDS THE KITCHEN

"Your ride's here."

"Call when you get there!"

"No, thank you. I'm happy with my current phone service."

I JUST BROKE UP WITH MY GIRLFRIEND. BOY! TALK ABOUT YOUR BIG BANG!
BeE
COMIC RADIATION

LIFE CYCLE OF A RICH AND FAMOUS SCIENTIST

PORTRAIT SPECIAL!!!

THREE 10X13'S

FOUR 8X10'S

100 WALLETS

10,000 MICROSCOPIC CUTIES THAT FIT ON THE HEAD OF A PIN.

1,000,000 KEEPSAKES CLONED RIGHT INTO YOUR DNA.

ZIE

"Do you have this in something a bit more endangered?"

"Well, is it Indian or African?"

"I know you adopted me, but who were my biological parents?"

"Nice little nest egg."

"She's in the henhouse awaiting the birth of our children."

"It's either a global warming trend or merely a fad."

"Another bird call?"

"Check this out! Reptiles colonize islands by hitchhiking on oceanic rafts."

THE MOTHER OF EVOLUTION

"I don't believe we've been formally introduced. I am Capra hircus, but everyone calls me Billy."

"It's been six years in the rainforest and it's getting harder and harder to claim a home office deduction."

Tell the wild animals I have arrived.
EGO TOURISM

"Now, will you admit we're lost?"

"How did you become interested in nature?"

"I can't go any farther. My cell phone will be out of range."

HI, EDDIE!
BLE

PROOF THAT EDDIE KNOWS HIS SNAILS

THE ENDANGERED
SPECIES ACT

"You're beaching yourself here?"

MAN: THE ONLY ANIMAL
TO PURPOSELY WEAR PLAID

BIOLOGY
CHEMISTRY
SCIENCE OF LITTLE ROUND THINGS

"Sorry I'm late. I volunteered for the obesity-gene study."

"Have you dined with us before?"

"Did you know that DNA spelled backwards is AND? Makes you think."

"Can you believe it? They're now cloning monkeys."

"Stop! Your grant is ending!"

"I've come for your exciting research."

"Stop being creative and discover something already."

"Mom, Dad . . . I just wanna thank you for your contribution of DNA."

GROUP PHOTO OF ETHICS SYMPOSIUM

"Sure, I got the huge staff, the huge salary, and the huge office, but . . . no view."

"We're a prestigious department, yet we have only three active prima donnas."

"One more upside-down slide and he can kiss tenure good-bye."

"Being a post-doc for thirty years can really screw up your life."

I MERELY WANTED TO READ BOOKS.
PARKING
FUNDING
PARENTS
SEMINARS
Industry
Agencies
LIBRARY
GOVERNMENT
OFFICE
REVIEWS
JOURNAL
YOU ARE HERE
FIELD WORK
labs
GRANTS
TALK
POST-DOC
PAPERS
RESEARCH
COMMITTEE WORK
COMMUNITY
STUDENTS
EXAMS
TA'S
PERSONNEL
TEACHING
ADMINISTRATION
Lectures
SPACE
ACCEPTANCE
REJECTION
ETHICS
AWARDS

IN THE NOT TOO DISTANT FUTURE
IN THE SORTA DISTANT FUTURE
IN THE WAY DISTANT FUTURE

"What difference does it make WHOSE fault it was?"

"I'm only here for breeding purposes, but I'm really enjoying myself."

"Gentlemen, I trust there won't be a problem working under a female bloodsucking mosquito."

*"About this medical marijuana study . . .
I'm totally stoked about the results."*

ANIMAL MAGNETISM

"Be tough. Never let them hear you squeak."

WHERE DO YOU GET YOUR IDEAS?

As a cartoonist I am frequently asked, "Where do you get your ideas?" Most ideas come from observing people, listening to conversations, noting frustrations, silly quirks and weird situations. Then the words or sketches go through a brainstorming session where ideas are exaggerated, given free association and word play, taken out of context, or twisted a little.

I worked in several research labs before I changed careers, so many cartoon ideas come from my past experiences. One conversation from a co-worker regarding administrators, "They oughta stick 'em all in one room and blow 'em the hell up" became a cartoon that later appeared in *The Wall Street Journal.* Another time, while I was searching for a library book an idea became "Science of Little Round Things" which first appeared in *American Scientist.*

On the following pages are sketches taken from my notebook. Here I'm a cartoonist on location. To the right of those pages are cartoons that resulted from these observations.

"Let me through! I'm a cartoonist!"

ARGUING OVER ERLENMEYER FLASKS.
ARE THEY HALF EMPTY or HALF FULL?

"You've heard of smart computers? Well, these are graduated cylinders."

TWO SCIENTISTS WORK
CLOSE TOGETHER.
WHAT ARE THEY STUDYING?
ARE THEY COLLABORATING?

ARE YOU WORKING ON THE MIRROR IMAGE OF MY COMPOUND?
YES!

4/18
10:07
Disaster strikes!
Total chaos!

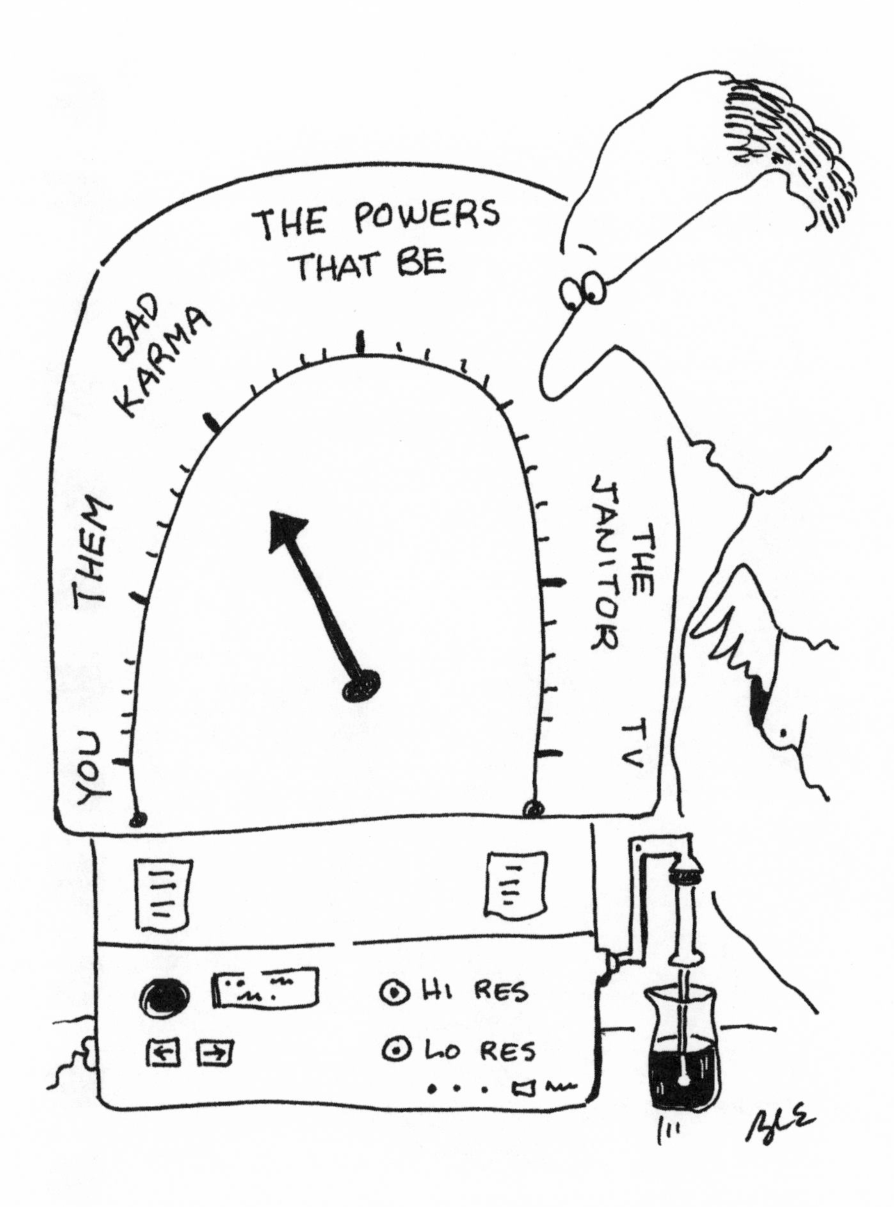
THE POWERS
THAT BE
BAD
KARMA
THEM
YOU
THE
JANITOR
TV
HI RES
LO RES

PONDERING THE LENGTH
OF SOME GUY'S SEMINAR.

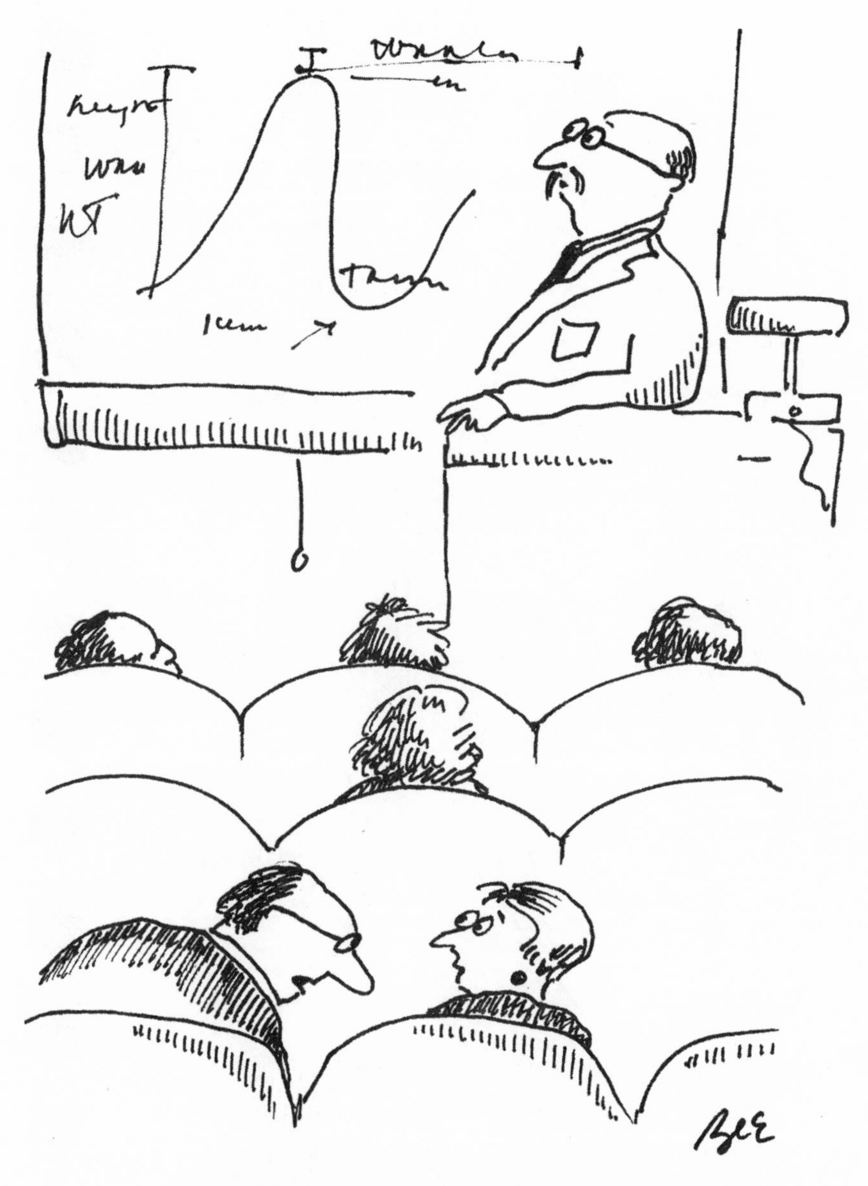

"Wake me if he mentions my stuff."

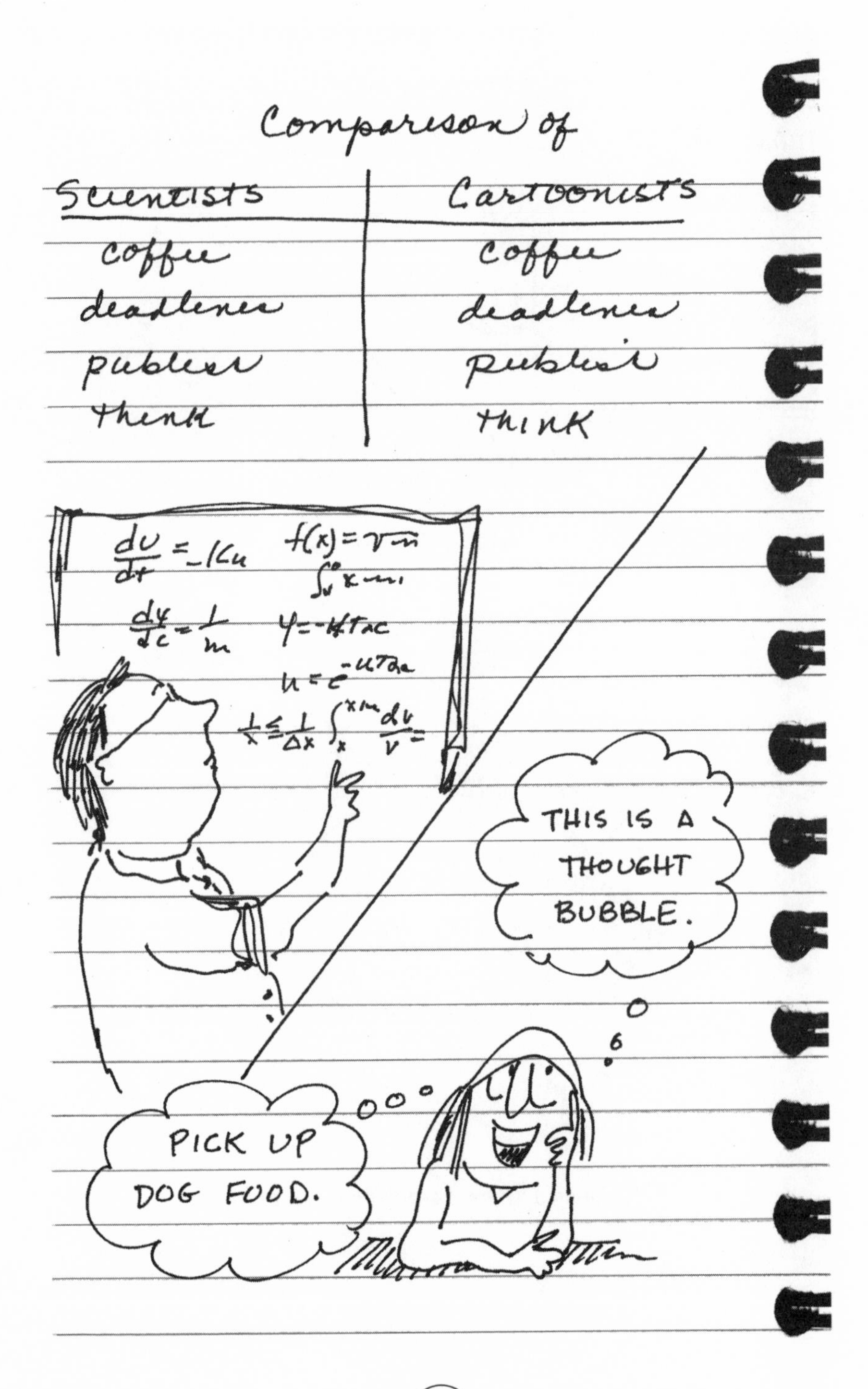
Comparison of
Scientists
Cartoonists
coffee
coffee
deadlines
deadlines
publish
publish
think
THINK
THIS IS A THOUGHT BUBBLE.
PICK UP DOG FOOD.

"Be still! There's a fine line between art and science."

"*If he knows so much about animal communication how come he doesn't know when to shut up?*"